# Table of Contents

**Ecosystems**..................................................................1
- What is an ecosystem?..............................................1
- What lives in an ecosystem?....................................1
- What are some unusual types of ecosystems?...........4

**Biomes**..........................................................................5

**Changes in Ecosystems**............................................6
- How do we know changes in ecosystems have occurred?......6
- Ecosystem and Climate Changes in Ancient Earth.....................7
- A History of Life...........................................................10
- Ecosystem Changes in Recent History........................12
- Ecosystem Changes Happening Today........................12

**Adaptations to Change**.............................................16

**Glossary**......................................................................19

# Ecosystems

## What is an ecosystem?

If you look out your window or walk into your backyard, you will be looking at an ecosystem. An **ecosystem** is a group of plants and animals that live together in a certain climate and specific landscape. All the living things in that ecosystem need the biotic (living) and abiotic (non-living) factors present in order to survive. An ecosystem can be as tiny as a mud puddle or as enormous as a rainforest. All of the parts of an ecosystem work together to keep everything in balance. A healthy ecosystem is balanced, but it is also always changing. Temperatures rise and fall. Animals and plants live, die, and decay. Fires and floods can cause changes in ecosystems.

If you look out your window or walk into your backyard, you will be looking at an ecosystem.

## What lives in an ecosystem?

In a balanced ecosystem each species is dependent in some way on every other species, usually through food chains and webs. A **food chain** shows who eats whom in an ecosystem. Food chains always start with a plant and end with either an animal or a fungus. A **food web** is similar, but instead of one organism eating one other organism, there can be many links with different animals eating different plants and animals.

pond algae

Think of a pond. There is algae in most ponds. The algae makes its own food through photosynthesis. **Photosynthesis** is the process used by green plants to convert energy from sunlight, water, and carbon dioxide into sugar and oxygen. Sugar, which is a carbohydrate, is used by the plant for nutrition and energy production. When a species makes its own food, like a green plant, it is called a **producer**. Producers produce their own food.

Small invertebrate animals like snails and minnows eat the algae.

Small invertebrate animals like snails and minnows eat the algae. Species that consume producers are called **primary consumers**. Small fish like perch eat the invertebrates. Species that consume primary consumers are called **secondary consumers**. The "secondary" means that that species is two steps removed from the producer in a food chain. Some ponds have larger fish like bass that eat the smaller fish. Sometimes aquatic birds consume the small fish, too. The bass and birds are called **tertiary consumers** or predators.

One food chain for our pond looks like this.

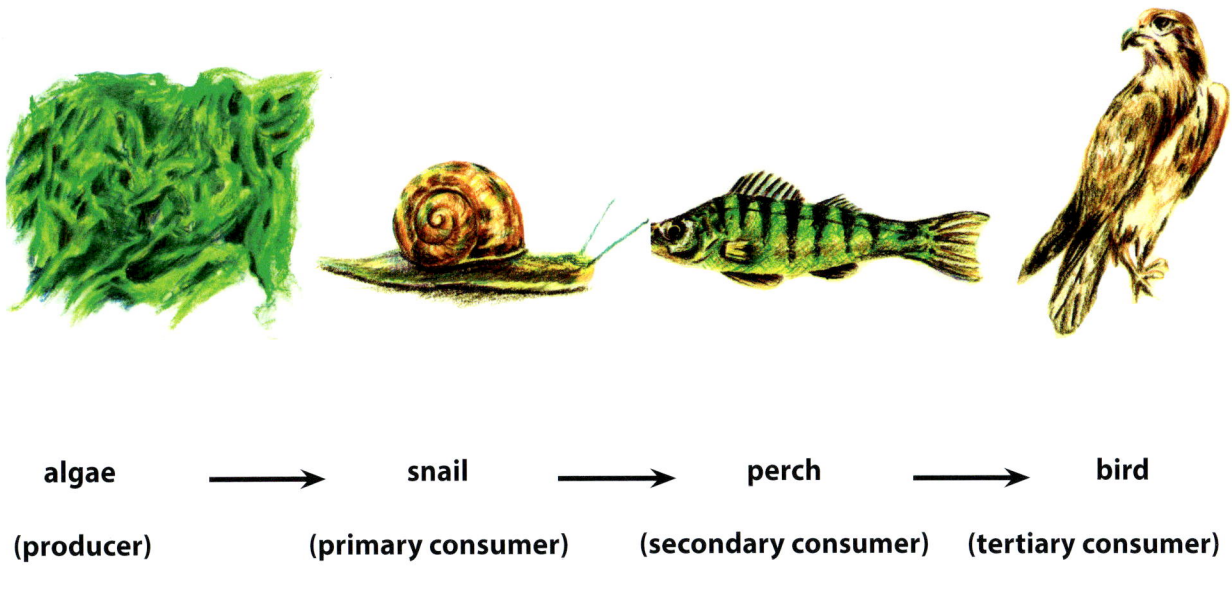

**algae** → **snail** → **perch** → **bird**

(producer)    (primary consumer)    (secondary consumer)    (tertiary consumer)

## Food Chain

**2**

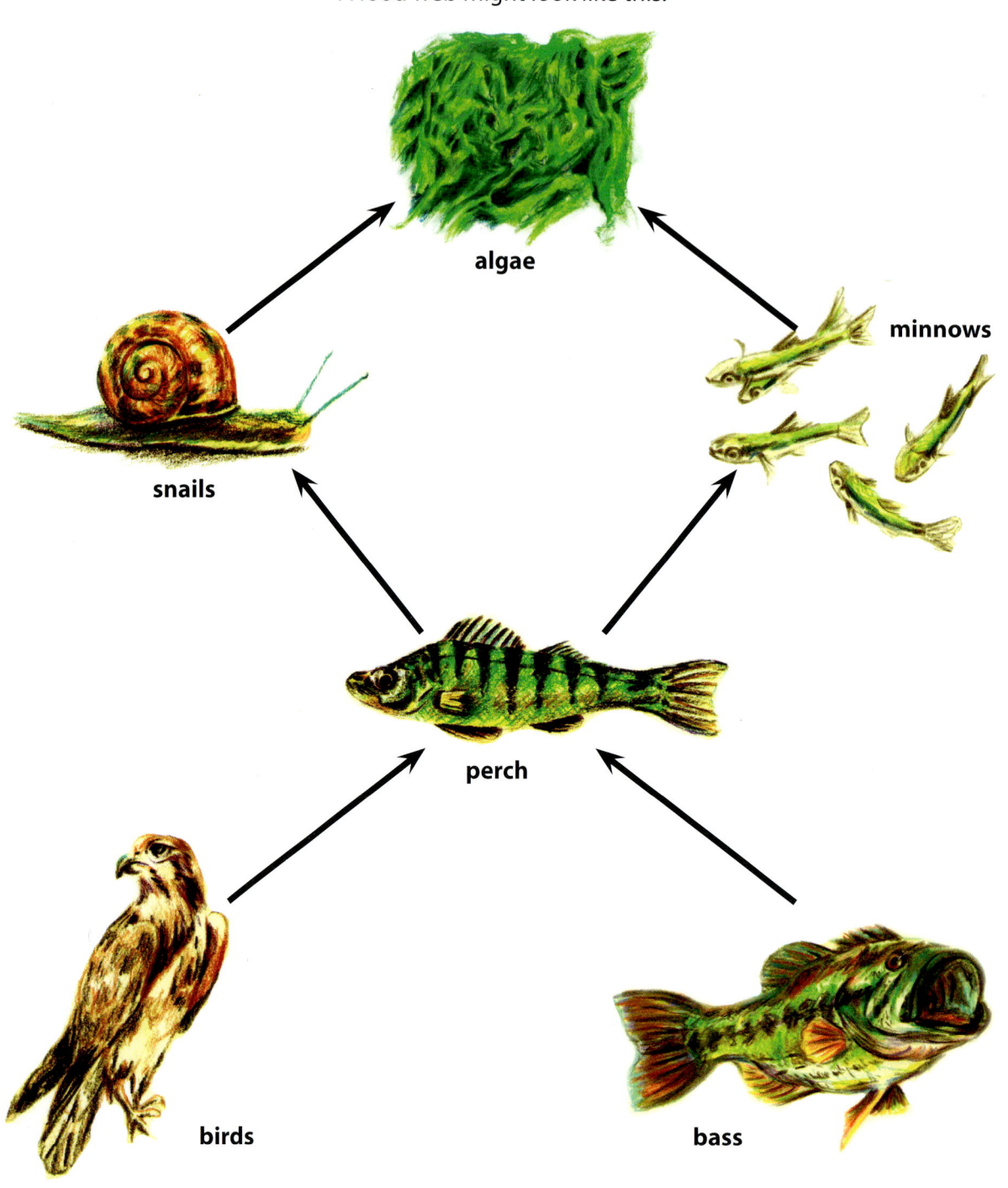

A food web might look like this.

**Food Web**

Note: Don't let the direction of the arrows confuse you. In food chains, the arrows point in the direction the nutrients move. When a snail eats the algae, the nutrients move from the algae to the snail. Because food webs can be more complex, the arrows point from the organism to what the organism eats.

When an ecosystem is in balance, the populations of different species will remain relatively stable. When changes occur that increase or decrease the population of one or more species in the food chain, the populations of the other species will change, too. Sometimes we can predict what those changes will be. Other times scientists don't discover the link between species until a change occurs and they observe a change in another population.

## What are some unusual types of ecosystems?

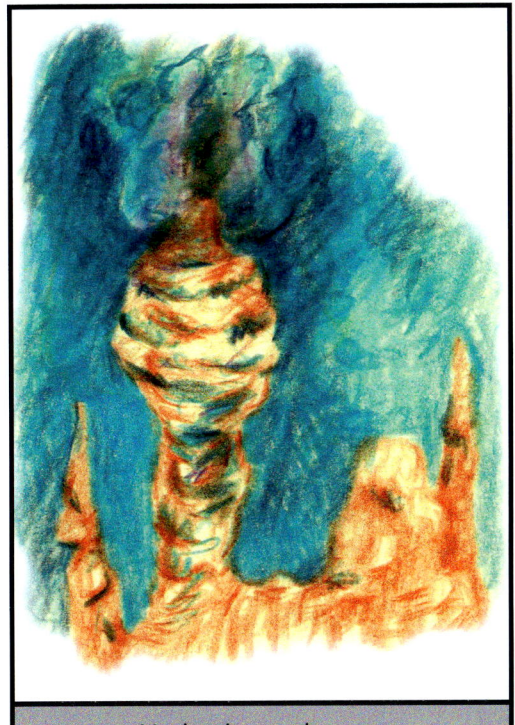

Hydrothermal vents

While most ecosystems are dependent on at least some green plants to be the producers in the food chains, there are a few ecosystems on Earth that have different or unusual producers.

**Hydrothermal vents** occur in deep ocean waters where tectonic plates are moving apart or breaks in the Earth's crust have occurred due to volcanic activity. The water at these vents becomes extremely hot, up to temperatures of 750°F. There is no sunlight able to penetrate to the depths of these vents. Scientists long believed no life could survive such inhospitable conditions. They were wrong. Red tube worms, giant clams, shrimp, and even octopi occupy these ecosystems. The producers are a type of bacteria able to convert methane or carbon dioxide into sugars using the sulfur compounds produced by the vents rather than sunlight.

At the bottom of some seas there are puddles of **brine**, extremely salty water that is much denser than regular sea water. The brine forms deep sea brine pools. No green plants are found at such deep places in the sea, around 2400 feet below sea level. Instead, there are mussels and bacteria that have formed a symbiotic relationship. A **symbiotic relationship** means that two species have a relationship that provides a benefit to at least one of them. The mussels and bacteria provide each other with all of the nutrition each needs.

Deep sea brine pools

**4**

Many of these brine pools coexist with cold seeps. **Cold seeps** are areas of the ocean floors where chemicals like hydrogen sulfide, methane, and other organic compounds ooze out from the ocean floor. The producers in cold seep ecosystems are usually bacteria that can convert the otherwise toxic chemicals emerging from the ocean floor into carbohydrates that other species can use as fuel. Cold seeps are found in most of the oceans on Earth.

Mussels and bacteria exist in brine pools and cold seeps.

# Biomes

If you were to group together all of the ecosystems that shared significant attributes like climate and similar plants or animals, that would be a biome. For example, there are deserts in Africa, Australia, Asia, South America, and North America. The desert in each location is a different ecosystem, but taken together, they make a biome. A **biome** is made of different parts

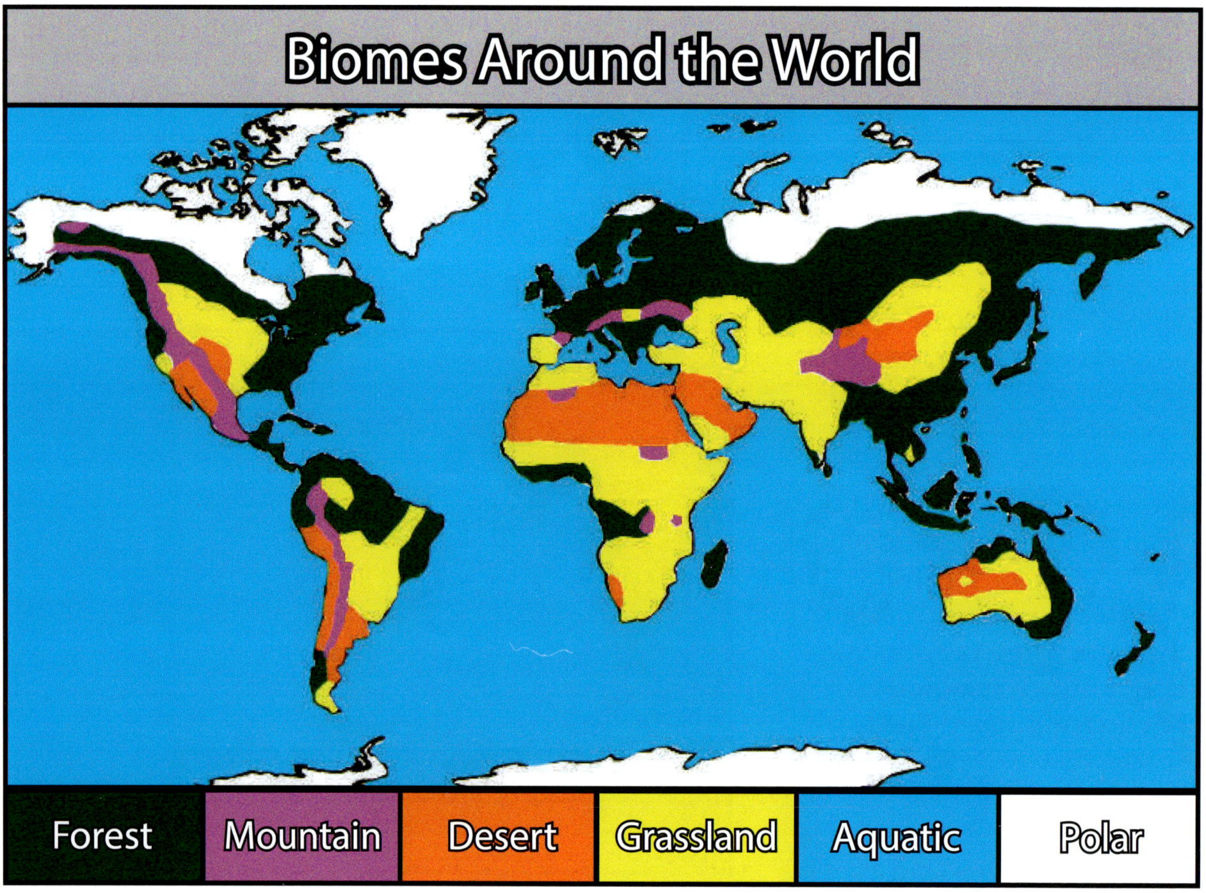

## Biomes Around the World

| Forest | Mountain | Desert | Grassland | Aquatic | Polar |

of the Earth's surface that share major climates, habitats, and living organisms. There are many ways to list biomes, but the major ones scientists use are desert, aquatic, grassland, mountain, polar, and forest. Each of these can be broken down based on different conditions. For example, there are temperate forests, taiga forests, and rainforests. Taiga forests tend to be colder than the other two. Rainforests, not surprisingly, tend to get more rainfall and humidity than temperate or taiga forests.

# Changes in Ecosystems

## How do we know changes in ecosystems have occurred?

One of the ways scientists can learn about ancient Earth is to study changes in its ecosystems. Climate change was one of the main drivers of ecosystem changes on the early Earth. Scientists have many ways to look for climate change in the distant past.

Remnants of pollen found in rocks or sediment can provide clues about the types of vegetation that were found in a certain area at a certain time. Since we know what climates most plants require for growth, we are able to figure out what climates existed at different times in the past.

If fossils of coral and fish are located, that indicates the area was underwater at some point.

Fossils can provide similar clues about the ecosystems in a given geographical location as well as for a specific timeframe. If scientists know where a fossil was found as well as the approximate date of the layer of rock or sediment in which it was found, they can draw conclusions regarding the ecosystem, climate, and conditions that existed at that place at that time. For example, if a layer of sediment contains many fossils of ferns, we know that location was, at one point, above sea level and probably in a temperate to rainy climate. If fossils of coral and fish are located, that indicates the area was underwater at one time.

Ice core samples

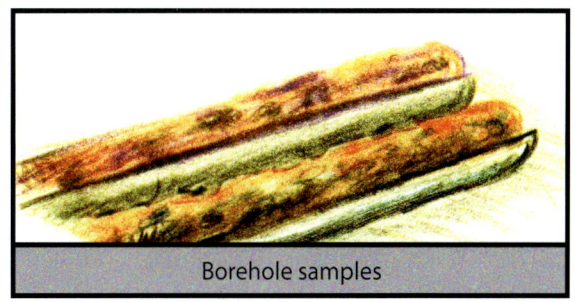
Borehole samples

Ice core and borehole samples give similar information. Ice core samples are long cylindrical rods of ice drilled from deep ice beds in Greenland and Antarctica. The longest ice core samples are almost two miles long. Borehole samples are similar, but rather than ice, the scientists drill out cylinders of rock and sediment from the Earth's crust.

In both cases, the concentration of carbon dioxide and other chemicals can give scientists much information about the makeup of atmospheres from long ago.

## Ecosystem and Climate Changes in Ancient Earth

When volcanoes erupt, they send enormous amounts of aerosols and volcanic gases into the air.

The Earth is and always has been undergoing constant change. These changes have determined which life forms flourished and which fell into extinction. There are many different factors that have caused the Earth to change over time.

### Volcanoes

When volcanoes erupt, they send enormous amounts of liquid and solid droplets (**aerosols**) into the air along with large volumes of volcanic gases. These aerosols can block sunlight from reaching the Earth's surface, thus lowering its temperature. This can cause widespread changes in climates around the world. Additionally, greenhouse gases in volcanic eruptions can change the makeup of Earth's atmosphere and impact plants and animals that are sensitive to carbon dioxide, sulfur, and methane levels. Scientists are able to tell when volcanic eruptions occurred by looking at the types of material found in boreholes. Lava becomes a certain type of rock, called igneous, when it cools. When large amounts of igneous rock are found, scientists know a volcano erupted. When the time scale of the rocks matches changes in carbon dioxide and sulfur levels found in ice samples of similar age, scientists are able to predict what effect that eruption may have had on the ecosystems nearby.

### Earth's Rotation

Changes in the Earth's rotation creates changes in worldwide weather patterns. The changes in

Earth's rotation work together to vary the amount of sunlight and heat that reaches the Earth from the Sun. Differences in sunlight cause differences in global climates. Knowing what the Earth's orbit was like at different points in the past allows us to draw conclusions about what the climate and ecosystems would have been like in different locations around the Earth.

## Atmospheric Changes

Life on Earth about 2.5 billion years ago was very different than it is today. One major difference is that organisms were anaerobic – they did not use oxygen. There was no oxygen in the atmosphere. Then the organisms living in the seas diversified, and some evolved the ability to perform photosynthesis. Remember that photosynthesis produces oxygen. This new gas, oxygen, had a catastrophic effect on ancient Earth. Oxygen was deadly to most life-forms. The level of oxygen rose significantly causing approximately 90% of all lifeforms to die. In this case, oxygen was a powerful form of pollution.

This catastrophic event, called the Oxygen Catastrophe, was a direct result of a change in a worldwide ecosystem. It was after these extinctions that oxygen-tolerant organisms began to form the complex life-forms we see today.

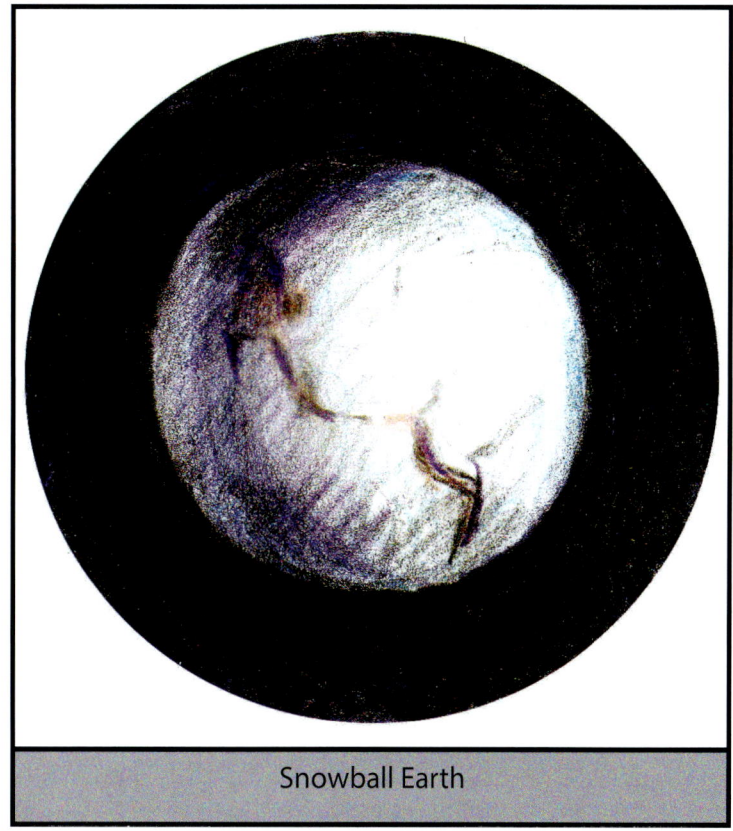
Snowball Earth

## Ice Ages

As oxygen levels rose, carbon dioxide and methane levels fell. This caused the average temperature on Earth to fall. Scientists also believe that the angle of the Earth allowed less direct sunlight to hit the surface. This would have lowered temperatures, too. By 2.1-2.4 billion years ago, large parts of the planet were covered in ice. This marked Earth's first ice age. Once again, large numbers of organisms died off. This time the cause was a lack of available sunlight passing through the ice sheets, depriving organisms dependent on photosynthesis of their ability to produce nutrients and energy.

The Earth has continued this cycle of heating and cooling. There have been at least four ice ages in the history of the earth, and we are still in one today!

Ice ages last millions of years. During an ice age there are times when the planet is dry, cold, and mostly covered in ice. These are called interglacial periods. Most ice ages have several glacial and interglacial periods.

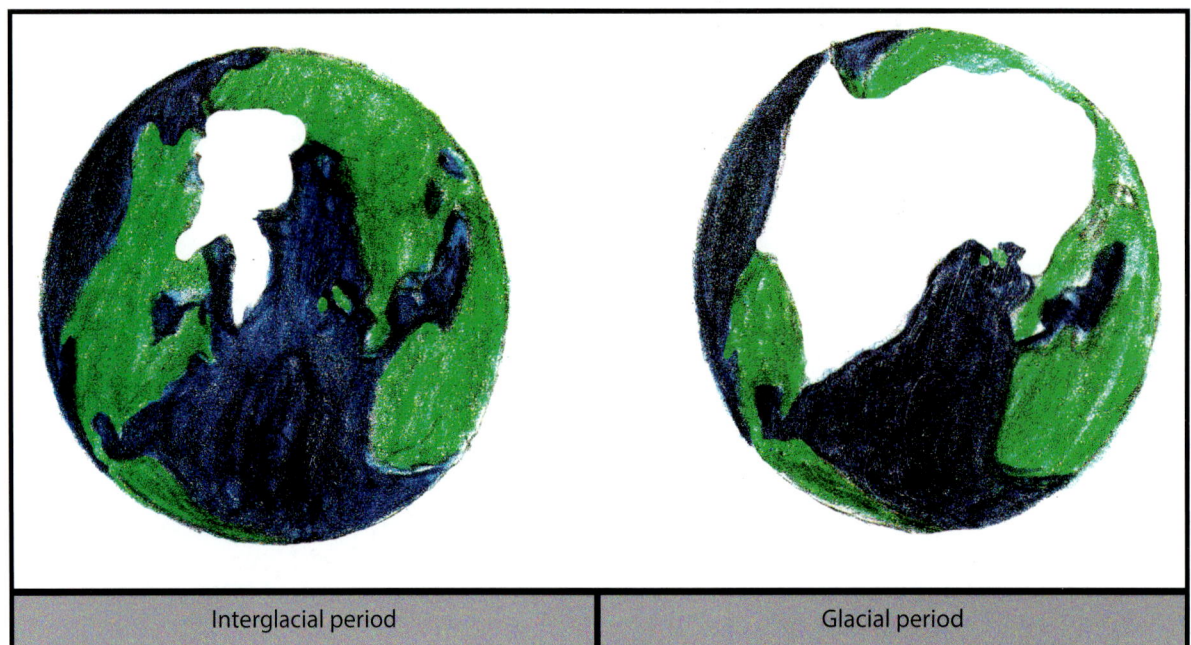

| Interglacial period | Glacial period |

During a glacial period, some ocean water is locked in great sheets of ice. This causes sea levels to drop. When sea levels drop, land bridges are exposed in areas normally covered by water. Animals migrate across these bridges, which close as the planet enters an interglacial period. As sea levels rise during the warmer interglacial period, the land bridges are once again covered by water. When the bridges close, the groups of animals on each side of the bridge are separated and begin to evolve separately.

One example of two groups of the same species evolving differently in different environments is the Sika Deer in Japan. The ancestors of today's deer migrated to the islands of Japan from mainland Asia during the late Pleistocene epoch, about 125,000 years ago. As sea levels rose and the land bridges were lost, the deer began to evolve differently. Today, each island has its own species of Sika Deer. There are significant genetic differences among the different island

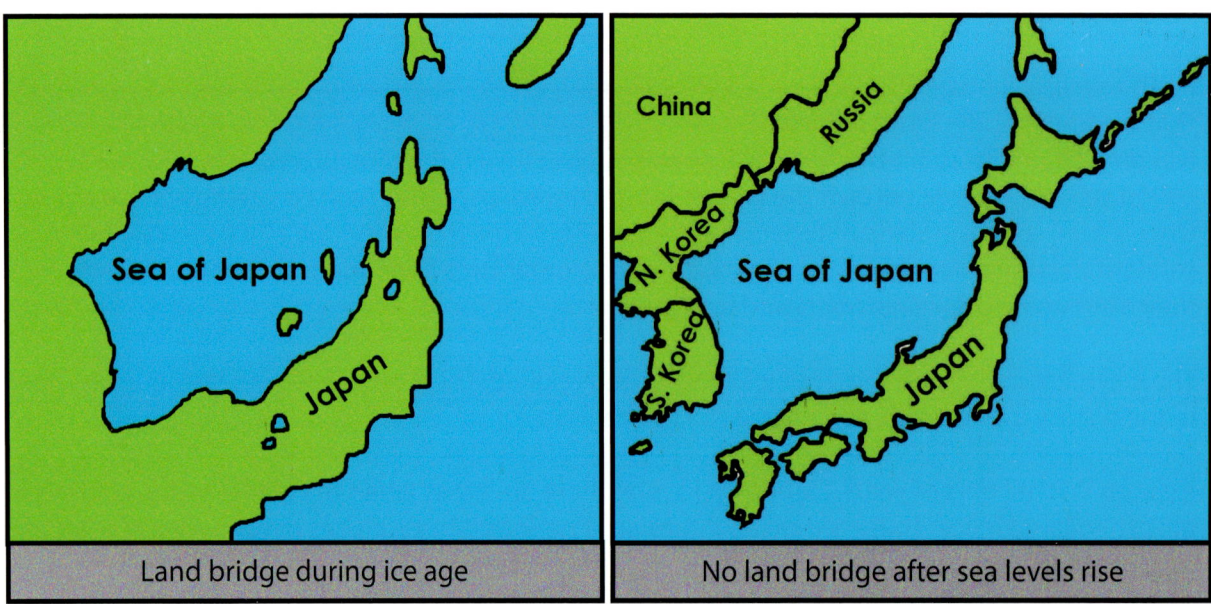

| Land bridge during ice age | No land bridge after sea levels rise |

Cambrian explosion

populations. Most notably, the deer from the northern islands are significantly larger than those from the southern islands. In some cases, the northern deer are twice as large.

### Other Factors
Ocean currents, movements of Earth's tectonic plates, changes in the orientation of the North and South poles, and astronomical events like meteorites and solar flares may all contribute to the cyclic cooling and heating our planet experiences.

## A History of Life

Very often there is a dramatic increase in the number and diversity of life-forms in the warmer periods following the ice ages. These are often called explosions.

The Cambrian explosion (542-488 million years ago) is the earliest era of massive numbers and types of life evolving following an ice age. During this age the first vertebrates appeared, and massive numbers and very different types of life-forms appeared. About 445 million years ago, during the Ordovician period, the seas were warm and shallow. Many marine species (organisms that live in the oceans) evolved and increased their numbers significantly.

During the Carboniferous period, about 350 million years ago, conifer trees (like pine trees), amphibians, winged insects, and huge freshwater fish appeared. As temperatures became more mild and movement of the tectonic plates created huge areas covered by land, animals developed the ability to lay eggs. This led to more animals venturing away from the waters to explore land.

The Jurassic period, 199-145 million years ago, was the Age of the Dinosaurs. Mild temperatures and a stable atmosphere allowed the giants like Diplodocus and Allosaurus to thrive.

Carboniferous period

The Cenozoic era, 66 million years ago to today, allowed the appearance of mammals and eventually humans. Stable temperatures and continent locations have allowed the ecosystems on Earth to remain balanced, allowing life to grow and flourish.

Jurassic period

## Extinctions
Changes to the environment have also led to times when when large numbers of organisms have died out at the same time. These are called extinctions. There have been 5 major extinctions.

About 443 million years ago the first mass extinction occurred. It is called the Ordovician-Silurian extinction. At this point, most life was found in the seas. Scientists believe that about 85% of Earth's life was wiped out because of changes to the environment, ocean levels, and the chemical makeup of the oceans during the ice age. Some theories suggest the Earth was hit by damaging radiation from space.

Over 80 millions years later, around 359 million years ago, the Late Devonian extinctions severely damaged the coral reefs in the oceans. One possible cause for this extinction is a comet or meteorite that could have hit the Earth with enough force to create earthquakes and shockwaves around the entire planet. Scientists think this event might have kicked up enough dust and debris to block out sunlight. This would have caused changes in climates across the Earth.

About 250 million years ago the Permian mass extinction occurred. This was the worst extinction event in Earth's history. Massive volcanic eruptions, changes in the atmosphere, increases in methane, and decreases in oxygen levels all may have contributed to this extinction event. About 96% of all living species died out during this time. This is the only time insects have ever been affected by an extinction event.

After the Permian mass extinction, about 201 million years ago, came the Triassic-Jurassic extinction. Climate change and volcanic eruptions are believed to have been the cause of this event. About half of all animal species died out, but not many plants were affected.

The fifth major extinction event occurred about 60 million years ago. The Cretaceous extinction saw the death of most of the dinosaurs. Massive volcanic eruptions changed the ecosystems around the world as well as vastly lowering sea levels.

Many scientists believe we are in the middle of Earth's 6th mass extinction event today. Much research is underway to determine if this is the case.

Clearly, changes in Earth's ecosystems have led to several significant changes in the plants and animals that inhabited ancient Earth. More recent changes have been observed, too.

## Ecosystem Changes in Recent History

The 10th-14th centuries, 900s AD to 1300s AD, are known as the Medieval Warm Period. During this time, warmer-than-average temperatures in both winter and summer increased harvests and led to a time of success in Europe. Plenty of available food gave people the chance to pursue exploration and artistic, architectural, and scientific development. Many of Europe's great cathedrals were built during this time of prosperity. Around 1300, temperatures began to drop. This era is known as the Little Ice Age. It led to environmental disasters, which caused failed crops and famine. Scientists think it might have played a role in how the Black Death was able to wipe out 25-50% of the population. Changes in ecosystems continue to impact the planet today.

## Ecosystem Changes Happening Today

Drought

Flooding

One example of a current cause of changing ecosystems is El Niño. El Niño is the name of an event that happens every 2-7 years. The waters in the Pacific Ocean become warmer. This upsets the weather systems in the tropics leading to flooding in some areas and droughts in others.

The sea level may change by over two feet. More tropical storms occur during El Niño, which can lead to massive destruction, erosion, and loss of crops. The causes of El Niño are not known, so predicting this event is difficult.

Fire can change the ecosystem in a matter of hours or days.

Fires, both man-made and wildfires, can change the ecosystem of enormous areas in a matter of hours or days. In 1988 a single lightning storm ignited several fires that burned almost 1.6 million acres in Yellowstone National Park. That is an area bigger than the entire state of Delaware.

Sometimes fire wipes out the native plants (and animals) in an ecosystem. Animals that can run and jump often outrun fires. Animals that are too small or slow often die in the flames. **Native plants** are plants that grow naturally in a given ecosystem. If the native plants are not fire-resistant, a fire might kill off the native plant communities and allow them to be replaced with introduced weeds. **Introduced plants**, often weeds, are plants brought into one ecosystem from another ecosystem.

More often, fires act to maintain the health of the plants in the community. Some organisms need fire to survive. The majestic giant redwoods in California take two years to develop their seed pods. These pods are sealed with a strong glue-like material. The pods stay on the trees until a forest fire heats them to the point that they pop open and disperse their seeds. Some wildflowers need the smoke from fires to allow their seeds to begin to grow. There is a species of wasp that lays its eggs only in the smoldering remains of forest fires. The eggs are deposited under the burnt bark of a cedar tree that is infected by a fungus that digests wood fiber.

Other organisms have **adapted**, or changed in response to their environment, to allow them to survive fires. Some shrubs produce new shoots that travel under the soil. This allows the plant to regenerate if the woody part of the shrub that is above the soil is lost to a fire. The ability

Giant redwoods in California

to adapt may mean the difference between existence and extinction for many plants.

Volcanic eruptions can have a catastrophic effect on ecosystems. In 1980 Mount St. Helens erupted with a terrible explosion. Hot gas and rock shot out of the volcano and traveled at over 670 mph to wipe out over 200 square miles of forests. Streams disappeared in the heat while large areas of snow cover melted and began flowing down the side of the volcano. The ecosystems around the volcano were permanently changed in an instant.

Much was lost, but life returned even to this scorched environment. Plant spores were carried into the area by winds. In the absence of competition from other plants, mosses and ferns began to grow in the area. Birds and insects flew into the area, too. Small creatures and seeds rode in on the feet or feathers of birds. The area around the volcano has not fully recovered, but new ecosystems have begun to thrive in most places.

It does not take a huge event to impact ecosystems. Even something as small as a microscopic virus or bacteria can wipe out a key plant or animal, causing an entire ecosystem to collapse. For example, there is a parasite that infects the brains of sea otters, causing death. In the Aleutian Islands, sea otters eat sea urchins. Sea urchins eat kelp. A major habitat of the Aleutian Islands is the undersea kelp forests that surround the area.

Kelp forest

In the 1990s the sea otter population fell to a low level due to deaths from the parasite. Too few otters eating sea urchins meant too many sea urchins survived and wiped out the kelp forest habitat. The loss of the kelp led to the loss of small fish in the area. Bald eagles, which eat the small fish in the kelp forests, changed their eating habits and began moving to different environments. The reduced photosynthesis by the kelp also caused a rise of carbon dioxide levels.

Unfortunately, human beings cause many negative changes to ecosystems all over the planet. Deforestation, pollution, over-hunting, illegally killing protected plants and animals, and even attempts to help ecosystems in trouble have all contributed to people's negative effects on habitats, climate, and ecosystems.

Macquarie Island is in the Pacific Ocean about halfway between Antarctica and New Zealand. In the 1800s, rats and mice jumped off of ships visiting the island. To control the rats and mice, sailors began bringing along cats. They also began putting rabbits on the island so there would be something for them to eat when they returned. The cats ate the rabbits and the cat population skyrocketed. Unfortunately, the cats also wiped out the population of two native species of flightless birds. Meanwhile, the rabbits continued to reproduce and ate huge tracts of vegetation leaving dry soil behind. In the 1970s, scientists introduced the deadly disease myxomatosis to bring the rabbit population down to manageable levels. This seemed to work and, for a while, the vegetation began to recover. With fewer rabbits to eat, the cats turned to species of burrowing birds for food. In 1985 scientists decided they needed to eliminate the cats to save the burrowing birds. By 2000 the last of the cats had been killed. But then, scientists realized the population of the rabbits was no longer being controlled by the myxomatosis, so the rabbit population was, once again, soaring. The rabbits ate so much vegetation that about 40% of the island was made barren, lacking any plant life. Erosion caused a landslide that almost wiped out a breeding population of King penguins. It took until April 2014 to eliminate the last of the pests from the island. It remains to be seen whether there will be continued, unexpected trouble with the ecosystem.

The cats ate the rabbits and the cat population skyrocketed.

# Adaptations to Change

Pandas are a good example of a species that is highly specialized.

When an ecosystem changes, survival of a species may depend on that species' ability to adapt to the new environment. Some species are able to adapt more easily than others. A species is considered a **specialized species** if it has evolved to thrive in a limited environment or to eat limited types of food. The more a species has limited itself to fit in a given environment, the less likely it will be that it will be able to adapt and survive when the ecosystem changes. Pandas are a good example of a species that is highly specialized. Pandas limit their diet to primarily bamboo. About 99% of their food intake is bamboo; they eat very small amounts of other food sources. If the ecosystem changes, say a fire or deforestation eliminates the bamboo sources, the pandas can, and do, suffer starvation. They are not able to adapt well. A species that is more generalized in its eating habits, for example an omnivorous black bear, is better able to adapt to the loss of one or more of its food sources. **Generalist species** are able to eat a variety of foods and thrive in different environments.

Plants can be specialists or generalists, too. Cacti are highly adapted to live in low humidity with high temperatures. An increase in annual rainfall would likely kill the cactus populations in an ecosystem. Weeds are notorious for being generalists. As any gardener who has tried to rid her lawn of dandelions can tell you, many weeds will thrive in almost any environment.

Problems in ecosystems can occur when an invasive generalist plant is introduced. Kudzu is native to Asia but was introduced in the United States in the 1800s. Today this wildly successful weed covers over 7 million acres of land in the southeastern United States. It crowds out native species of plants and removes food sources for many primary consumers. This has led to serious disruptions of ecosystems all over the Southeast. Some people call kudzu "the vine that ate the South."

If an animal is able to adapt to the behavior of human beings, its chances of surviving in human habitats are greatly increased. Rats and mice are two examples. Both rodents can eat almost anything. They live anywhere people live by picking food out of garbage cans, eating food stuffs found in homes or silos, or stealing from pets. Brown rats are found on every continent except Antarctica. Their ability to thrive in almost any environment has allowed them to adapt to a wide variety of ecosystems.

Cockroaches are the most famous adapters. They are ancient organisms that have lived in almost every habitat on Earth since over 300 million years ago.

The ultimate survivors are those that have been able to adapt to the most extreme conditions. These organisms are called **extremophiles**. The plants and animals found in cold seeps and near thermal vents are all extremophiles.

Ecosystems change all over Earth for a variety of reasons. For an ecosystem to be healthy, each member must be present at a relatively stable population level. Any disruption to any member of an ecosystem can lead to a collapse of the entire community. The more adaptable a species is, the better it will be able to survive changes when they come.

# Glossary

**adaptation** - change in response to a change in the ecosystem

**aerosol** - liquid and solid droplets in the air

**biome** - different parts of the Earth's surface that share major climates, habitats, and living organisms

**brine** - extremely salty water, 3-5 times saltier than regular sea water

**cold seeps** - areas of the ocean floors where chemicals like hydrogen sulfide, methane, and other organic compounds ooze out from the ocean floor

**ecosystem** - a group of plants and animals that live together in a certain climate and specific landscape

**extremophiles** - organisms that exist in unfriendly environments, which can include high temperature, high pressure, lack of oxygen, or lack of light

**food chain** - a representation of who eats whom in an ecosystem

**food web** - similar to a food chain but with more options for the organisms to eat or be eaten by many members of the ecosystem

**generalist species** - organisms able to eat a variety of foods and thrive in different environments

**hydrothermal vents** - places in the deep sea where extremely hot water issues out of breaks in the Earth's crust

**introduced plants** - plants brought into one ecosystem from another ecosystem